The
Creative
Code

How Quantum Physics Unlocks Our Brain's Potential

Kendir Ramiz

The Creative Code: How Quantum Physics Unlocks Our Brain's Potential

INTRODUCTION

Here we are. As you read these words, you've probably never given the term "quantum" so much thought. Perhaps the word conjures images of complex formulas, lab coats, and a world far removed from your everyday life. You might even be right. But let me tell you something; I used to think the same way—until I took a step into this enigmatic world.

This book is a reflection of my personal journey—a curious mind trying to uncover the invisible connections between our brains and the universe. Quantum physics, at first glance, may seem like a complex field only accessible to scientists. But in reality, its rules might be deeply intertwined with our inner world – our minds. In this journey, with the guidance of both science and our own experiences, we'll attempt to reveal these connections.

Kendir Ramiz

Remember, every journey begins with a single step, and that step can be taken anywhere a curious heart beats.

Perhaps the most important tool in this journey is questioning. To question what we know and what we don't know in every moment of our lives, like a question mark. Because there isn't such a thing as 'accepting everything as it is'. In this book, rather than giving you the answers, I aim to make you ask questions.

Why do we think this way? Why do we behave that way? What do we really know, and how much of it is unknown? While quantum physics might not provide concrete answers, it does give us the opportunity to change the way we think and to gain new perspectives. We'll evaluate everything we know and don't know, push the boundaries, and set sail for new discoveries together.

If you've come this far, you're probably someone who doesn't want to be limited, just like me, right? Human potential is as boundless and undiscovered as the universe. In this book, by exploring the connections

The Creative Code: How Quantum Physics Unlocks Our Brain's Potential

between quantum physics and how our brains function, we'll look for clues to unlock this potential.

Maybe creativity isn't a gift some people have but is a latent power inside all of us, waiting to be awakened. Perhaps the mysteries of the quantum world hold the keys to unlocking this power. Maybe, each one of us, is a unique, creative being surpassing all boundaries. This book will be a journey in pursuit of these 'maybes'.

"Quantum," as I said, is a term that can be difficult to understand for many people. But don't worry; I won't bog you down with formulas and complex mathematical expressions. Instead, I'll try to explain the basic concepts of the quantum world in an accessible language, with examples from our daily lives.

Superposition, entanglement, quantum uncertainty…These words might sound foreign to you at first, but trust me, when you understand them, you'll gain a completely new perspective on how the universe and our brains work. It'll be as though a door is opening to an invisible world, and as you step through that door, everything will take on a different meaning.

Quantum physics is a branch of science that causes us to question everything we know. And maybe that's its most beautiful aspect. To discover that the universe is far more peculiar, complex and mysterious than we imagined... Perhaps what we call 'reality' is just a reflection of our own perceptions.

In this book, we'll step outside the ordinary, we'll move away from our comfort zones, and we'll encounter new ideas. Maybe, by the end of this journey, we'll find the opportunity to understand ourselves and the world better.

Quantum physics can offer some clues about how our brains function. Maybe our thoughts, emotions, and consciousness are closely connected to the rules of the quantum world. When we unravel these connections, we can understand human potential on a deeper level.

The Creative Code: How Quantum Physics Unlocks Our Brain's Potential

For me, this is not just a scientific discovery, but also a journey of inner discovery. And I invite you to join me on this journey.

Creativity has always been a subject of wonder throughout human history. Artists, scientists, inventors…all have changed the world by using their creativity. So what is the source of creativity? Where does this inspiration come from?

Neuroscience shows that creativity is connected to neural networks in the brain. But perhaps quantum effects play a role in how these networks operate. Maybe creativity, like the quantum world, is a bit mysterious, a bit peculiar.

Maybe, all of us have limitless creative potential. Maybe, the key to unlock this potential is in understanding how our brains function. Quantum physics can offer new perspectives in this respect.

In this book, we will try to discover how creativity blooms in our minds and what role quantum effects play in this process. Perhaps we can awaken the creative within us.

Quantum physics allows us to break free from our thought patterns. Maybe, in order to improve our creativity, we should reshape our thoughts with the rules of the quantum world.

Perhaps the key to creativity is to overcome our limits of thought. And quantum physics can offer us this key. This book will be a space where we can break our thought patterns, discover new ways, and unleash our creativity.

The Creative Code: How Quantum Physics Unlocks Our Brain's Potential

The Quantum Genesis of Computation

The story of computation, in many ways, is a mirror reflecting our own intellectual journey—a constant attempt to understand and control the world around us. For centuries, we've been captivated by the idea of building tools that can process information, solve complex problems, and even mimic our own thought processes. This ambition led us to classical computing, built on the seemingly simple language of bits – those fundamental 0s and 1s that have shaped the digital world as we know it. These bits, acting as switches that are either on or off, are the building blocks of everything from the smartphone in your pocket to the massive data centers that power the internet. It's a logic that is straightforward, reliable, and yet, in a very real sense, limiting.

The limitations of this classical approach become particularly obvious when we venture into mimicking the complexities of the natural world. Imagine trying to simulate the intricate dance of molecules during a chemical reaction, or the folding process of a protein—a process that holds the key to so much of the biology around us. These aren't simple binary choices; they're dynamic systems where possibilities abound simultaneously, where the simple on or off of a bit is not nearly enough. It is as if we have been using a hammer when we really need a scalpel and a telescope. Our classical computers, with all their power, have come up against a very real wall.

And it's this very limitation that brings us to the realm of quantum computing, a completely different kind of calculation that is based on the bizarre, counterintuitive, and incredibly beautiful laws that govern the subatomic world. Quantum computing isn't simply an extension of our classical machines; it's a departure—a way to compute based on the underlying rules of our reality itself. This is where we start to encounter the strange

The Creative Code: How Quantum Physics Unlocks Our Brain's Potential

and, to be honest, somewhat mind-bending concepts of the quantum world.

At the core of this revolution lies the qubit, the quantum bit. Unlike its classical counterpart, the qubit isn't restricted to being either a 0 or a 1. It can exist in a state of superposition, where it is both a 0 and a 1 simultaneously. It is as if you have a coin spinning in the air, not yet heads nor tails, but both at the same time. This might sound bizarre and counterintuitive, and you wouldn't be wrong for feeling this way. This superposition isn't a quirk of the subatomic world; it is the very source of the power that quantum computers are meant to harness. Instead of working through possible solutions one after another, as classical computers do, a quantum computer, by virtue of this state, can explore multiple possibilities simultaneously. This dramatically speeds up calculations, allowing us to tackle problems that are intractable for even the most powerful classical computers.

But how does this weird state work? It's all based on the fundamental principles of quantum mechanics, a theory that governs the realm of atoms and subatomic particles. It is a realm where the rules of our everyday experiences break down, replaced by a set of rules that can seem contradictory and even fantastical. It was this very realm that inspired the great physicist Richard Feynman to propose, back in the 1980s, that we need to build machines that would simulate quantum systems using quantum principles. He realized that if nature operates using quantum mechanics, then the most efficient way to understand nature would be to create computers that use quantum mechanics themselves. It was a revolutionary insight, and it laid the foundations for what we are now starting to realize. (Reference: Feynman, R. P. (1982). Simulating physics with computers.)

Another very strange aspect of the quantum world that quantum computers are looking to harness is entanglement. Entangled qubits are like two coins that are somehow correlated in a very strange way. When you measure the state of one, you instantly know the state of the other, no matter how far apart they are. It's not because of a physical connection or that information

The Creative Code: How Quantum Physics Unlocks Our Brain's Potential

is traveling faster than the speed of light. It's the very nature of the way that these two qubits exist and are described, as one. Einstein himself famously called this "spooky action at a distance," struggling with the idea that measuring one object could instantaneously affect another, no matter how far away it is. Yet, despite his discomfort, it is now a well-established part of the quantum world, and one that we can use to our advantage.

The actual process of constructing a quantum computer is, in reality, an incredible challenge, requiring a degree of precision that is almost impossible for our human minds to comprehend. These machines have to be cooled to temperatures colder than outer space and are susceptible to tiny disturbances and fluctuations in their environment, leading to errors that can make their computations meaningless. Imagine trying to build a perfectly balanced sand castle on a beach where every wave could destroy it, and you'll have a picture of the challenge that these scientists are going through. Despite these challenges, progress continues. Technological giants such as Google, IBM, and a great number of research groups are all working tirelessly to make the idea of a fully functioning quantum computer a reality.

Are we on the verge of a new era where our old computers become obsolete? Not quite, at least not in the way you might imagine. For the everyday tasks that we face, such as browsing the internet, writing a document, or watching a video, our current classical computers will remain perfectly suitable. In the coming years, the real promise of quantum computers will be in those areas that need very high computational power, such as the development of new medicines, the discovery of new materials, the optimization of complex systems, and so on. And it's within this realm of great possibilities, that the potential connections between quantum mechanics and our own creative process comes into full view. It's here that our journey to understanding the creative code within our brains will truly begin.

The Creative Code: How Quantum Physics Unlocks Our Brain's Potential

Echoes of the Quantum in the Mind

The idea that the human brain, that incredibly complex and mysterious organ that resides within our skulls, might have a quantum dimension to it, is a thought that has captivated my mind, and I suspect, the minds of many others throughout history. For centuries, we've tried to understand the inner workings of the brain, using the tools of biology, psychology, and neuroscience. We've mapped neural pathways, identified neurotransmitters, and categorized various cognitive functions. Yet, despite all this progress, there remains a feeling that we have not grasped the full picture, that some essential piece of the puzzle is still missing. It's as if we have described a beautiful symphony by only noting down all the instruments and their locations within an orchestra. We are still looking for that key that can unlock our understanding of how these elements give rise to something as amazing and mysterious as consciousness and the experience of being a human.

Kendir Ramiz

This brings us to the question that has been nagging me for a while now: if the universe itself operates on quantum principles, could it be that our brains, which are very much a part of the universe, also exhibit some quantum behavior? It's a question that might seem far-fetched at first, perhaps even a little too fantastical. But when you consider the strangeness of the quantum realm and the still-unsolved mysteries of the brain, it doesn't seem unreasonable to consider. For now, the evidence is suggestive, and perhaps circumstantial. But I think the connections are too intriguing to simply dismiss.

The human brain, this magnificent network of approximately 86 billion neurons, each connected to thousands of other neurons, is a system that dwarfs in complexity anything we have been able to create ourselves. (It is fascinating to think about the number of human minds that it took to create even a simple calculator, but all those human minds combined would never come close to the amount of data processed by just one single human mind in a fraction of a second).

The Creative Code: How Quantum Physics Unlocks Our Brain's Potential

These neurons, firing electrical and chemical signals, create the basis of our thoughts, emotions, memories, and our very sense of self. Yet, how do these myriad interactions, these tiny sparks of activity, give rise to our experience of the world, our consciousness, and our ability to create? These questions have challenged scientists and philosophers alike for centuries, and it's where the idea of a quantum brain begins to look like a possibility.

Our traditional view of the brain, much like our view of the universe before the discovery of quantum mechanics, is rooted in the classical, the deterministic, and the predictable. We've viewed the brain as a kind of biological computer, a complex machine, processing information in a way that is similar to, but far more complex, than the digital computers we are so used to. This classical view may indeed be valid when it comes to explaining a certain range of processes in the brain. However, it might be this limited viewpoint that's holding us back from understanding the brain's most unique and mysterious abilities. It is as if we have been trying to understand the vastness of the ocean by only looking at a small pond.

Perhaps we've been too quick to dismiss the possibility that quantum phenomena, which are now known to be a key part of nature at its most fundamental level, may play a role within the workings of the brain. It isn't that we want to abandon all that we have learned, all of that solid data. However, it might be that the mysteries of our minds, the very source of our creativity, require us to go further and explore these new possibilities, to examine our brain in a different kind of light. One area where we have actually seen quantum phenomena influencing biological processes is photosynthesis. Plants, as we know, use the energy of sunlight to create the sugars that allow them to live, and what scientists have discovered is that they do it with a process that makes use of quantum superposition.

Plants utilize the quantum effect of superposition to explore all potential energy pathways simultaneously, maximizing the process of converting light into chemical energy. (Reference: Engel, G. S., et al. (2007). Evidence

The Creative Code: How Quantum Physics Unlocks Our Brain's Potential

for wavelike energy transfer through quantum coherence in photosynthetic systems.) This incredible efficiency is made possible by quantum principles. Think about this for a moment, even plants use this bizarre, strange, yet incredibly useful principle in nature. If plants can use quantum effects, then why not the brain? Perhaps we have simply not yet found the evidence to show how and when the brain uses them? Or perhaps we are simply not seeing the quantum effects because our tools and models are simply too limited by our classical, deterministic view of the world?

If plants, in their silent way, are using superposition, could our brains be doing something similar in their complex operations? The connections between neurons, the constant dance of electrical signals, the interplay of neurotransmitters; it is perhaps too quick of us to assume that these interactions are purely classical in nature. Our current tools of neuroscience, while powerful, are themselves limited, and we may just be missing that essential piece of the puzzle. The very act of consciousness itself, that sense of being 'here' in the world, has resisted every attempt at being understood from a purely classical perspective. It has remained one of the most challenging questions for science. Perhaps the reason why we are struggling is simply because we are not using the right tools. Maybe we need to include and incorporate the principles of the quantum world to fully comprehend how our brains work.

And it's not just about consciousness. What about creativity? The ability to generate new ideas, to create art, to invent new technologies, these all come from an ability to think in novel ways. These moments of creativity, these sudden bursts of insight, do they arise from a purely classical brain? Or are they in some way, influenced by the more subtle, almost invisible, laws of quantum physics? It is in these moments of sudden, unpredictable, and seemingly spontaneous ideas that I find myself wondering about the connections between quantum mechanics and the human mind.

Perhaps our brains, in those moments of deep thinking and creativity, are acting a little bit like a quantum computer, exploring multiple possibilities simultaneously, before settling into a single, novel solution. Maybe this is

The Creative Code: How Quantum Physics Unlocks Our Brain's Potential

why creative insights often seem to come from nowhere, from the unexpected, from the very edges of our consciousness. They feel strangely random, almost like a flip of a coin, but in reality, there might be more than meets the eye. These ideas might seem far-fetched, of course. I get it.

Science, however, has always been about exploring the unknown. It's about questioning our assumptions, looking for different angles, and daring to challenge the status quo. The idea of a quantum brain, then, isn't about denying everything we've learned about the brain so far. Rather, it's about seeing if the principles of quantum mechanics can offer new and potentially powerful tools and ideas for understanding some of the brain's most profound mysteries, and for possibly understanding the source of human creativity itself. It is through this exploration that I believe, we can truly begin to unlock the nature of the human mind, and even our place in the universe itself.

Kendir Ramiz

The Creative Code – Quantum Explorations

The idea that the human brain might harbor a quantum dimension isn't just an academic exercise, a theoretical musing for late-night discussions in quiet libraries. It's an idea that becomes particularly compelling when we start to grapple with the nature of creativity itself. The brain, as we've established, is not simply a passive receiver of information. It's also a powerhouse of imagination, a generator of novel ideas, a canvas where the human spirit expresses its deepest longings and visions. I find myself continually returning to this idea: could it be that the unpredictable, surprising, almost magical quality of creativity is, at least in part, a result of the brain exploring a spectrum of possibilities through the lens of quantum mechanics?

The Creative Code: How Quantum Physics Unlocks Our Brain's Potential

Think about those moments of sudden inspiration, those flashes of insight that often appear to come out of nowhere. You might be walking down the street, or taking a shower, or just sitting in silence, and then, suddenly, a new idea pops into your mind, seemingly fully formed. This experience is something that most people have had, and it's always seemed like something that is beyond our normal human understanding. Traditional neuroscience tends to describe these events as a result of neurons firing in new and unique patterns. Different regions of the brain begin to communicate in unique ways, new connections are established and previous ones are strengthened. It is as if a unique, novel symphony is being played. While this classical explanation is valid for a lot of the brain's activity, we must allow ourselves to explore if there are quantum elements at play in these moments.

However, what if there is more to these moments than just neurons firing? What if our brains, at some subtle, underlying level, are also acting as quantum systems, exploring multiple possibilities simultaneously, much like qubits in a quantum computer, before converging onto a single, novel solution? Just as a quantum system can exist in multiple states at once, before collapsing into one state, perhaps our creative process also uses a similar approach, exploring ideas and concepts in many different ways, until a clear solution emerges. And what if some of these possibilities, those that lead to the greatest creativity, are the ones that are most influenced by quantum fluctuations and quantum probabilities?

The very act of creation itself seems so incredibly nonlinear, unlike the more predictable actions and thought processes that take up most of our daily lives. We might have a problem in front of us, try to solve it one way, and then another. But then, out of nowhere, comes an idea that solves the problem in ways we couldn't have previously imagined. It's as if, in these moments, our minds are making use of a different kind of logic, one that is more fluid, more unexpected, more akin to the strange world of quantum mechanics.

Maybe, during these moments, our brains don't just look at one possibility, one thought at a time. Perhaps the brain uses multiple paths of thought simultaneously,

The Creative Code: How Quantum Physics Unlocks Our Brain's Potential

exploring ideas and concepts in many different ways at the same time, until a clear and often unexpected solution or creation appears. And perhaps, it is this quantum capability of simultaneously considering many possibilities, that explains the human ability to generate such incredible diversity in art, science, and technology.

I find myself wondering if perhaps the most striking parallel between the quantum world and our creative process, is the crucial role of uncertainty. The principle of quantum mechanics is that there are very real and fundamental limits to what we can know at any one point in time. The Heisenberg uncertainty principle, which describes how the position and momentum of a subatomic particle cannot be determined with great precision, is a classic example of this. What if this concept of uncertainty, of the inherent limits to what we know, is not just a physical principle, but also a key component in our creative process?

Could it be that the creative process only becomes possible when we allow our minds to operate in a realm of uncertainty? That creativity comes from our ability to go beyond the status quo, to explore new, uncharted territories, and to challenge what we already know? Could it be that our moments of greatest insights come when our minds let go of the desire for total certainty, and instead, allow themselves to explore the unknown, to allow new possibilities to emerge?

This makes me think of those moments where artists, when confronted with a blank canvas, don't always have a clear plan, or a fully formed image of what they want to create. Instead, they allow the canvas, the colors, and the very act of creating itself to guide them, letting the artwork emerge, almost as if by itself. And I can't help but wonder if they are also tapping into a kind of quantum uncertainty that allows them to break free of their own expectations, allowing the creative potential of their brain to blossom. Perhaps, we are only creative when we let our brains, and our minds, to operate in the space of uncertainty.

When you think of the history of science and technology, it is often those individuals who have dared to question the fundamental principles of their time, who have gone

The Creative Code: How Quantum Physics Unlocks Our Brain's Potential

on to make the most important discoveries. These were scientists and engineers who were not afraid of uncertainty, but rather used their willingness to explore the unknown to fuel their curiosity.

These questions and insights lead me to consider if there is a hidden code, an underlying structure, to our creativity. We know that DNA contains the code for our biological structure. But what if there's also a hidden code for our minds, our imagination, and our creativity? And if that's the case, could it be that this hidden code is in some way connected to the strange, counterintuitive, yet deeply beautiful, world of quantum mechanics?

The idea that quantum principles might be at play in our creative process is, of course, highly speculative. And while we don't have the answers, it does give us a new lens through which to examine some of the long-standing questions that we have regarding how our brains work, how our creativity emerges, and the very nature of our consciousness.

This exploration also leads me to wonder if a deeper understanding of our brain's potential also requires a more intimate familiarity with the quantum world. Perhaps to unleash our own creative potential, we need to have a better understanding of the quantum realm, and use these insights to change the way that we see ourselves, and the way we think about how our minds work.

We are at the start of a journey that could very well change the way we see ourselves. But as with all great journeys, there is a degree of risk, but more importantly, an incredible amount of possibility and potential that could greatly transform how we view ourselves, and the world around us. It is in this context, that we must allow ourselves to explore these new possibilities with a sense of wonder, and hopefully, along the way, discover some truly remarkable truths about ourselves, our potential, and the very nature of reality itself. (Reference: Kounios, J., & Beeman, M. (2014). The cognitive neuroscience of insight.)

The Creative Code: How Quantum Physics Unlocks Our Brain's Potential

It is in this context, that the very act of exploring the quantum nature of our minds, may itself become an act of creativity, as it opens us to new possibilities, and new ways of seeing ourselves and our creative potential.

Kendir Ramiz

The Quantum Symphony of the Senses

The exploration of the quantum brain isn't merely an abstract intellectual exercise; it's a journey that fundamentally challenges our understanding of what it means to be human, how we experience the world, and where our creative spark originates. Up until this point, we have explored the possibility that quantum principles might be playing a crucial role in the brain, allowing our minds to perform computations that are not possible under the rules of classical physics. But these are mostly theoretical considerations. But what about our actual experience? How does our experience of reality itself interact with the quantum realm? These are questions that go far beyond our usual scientific and philosophical concerns.

The Creative Code: How Quantum Physics Unlocks Our Brain's Potential

It's tempting to think of the senses as simple input devices, gathering data from the world, which is then processed by the brain. But this view, while convenient, seems almost reductionist when you consider the sheer complexity of our sensory experiences. Consider the experience of listening to music, or viewing a piece of art, or tasting a meal made by a loved one. These experiences are not merely a collection of data. They are, in essence, a symphony of interconnected and interrelated elements that come together to create a rich and immersive feeling, and it's these rich feelings, that we consider the essence of being human.

The brain doesn't simply process sensory data; it actively interprets it, shapes it, and weaves it together with our emotions, our memories, and our expectations to create a unique subjective reality. What is true to one person might not be true for another, as each of our brains creates a different kind of personal experience. As the author Anaïs Nin beautifully put it, "We don't see things as they are, we see them as we are." This quote perfectly encapsulates how our inner state shapes our perception of reality. It's this deeply subjective element of experience that often makes it hard to see our reality from the perspective of another individual. The very act of sensing is not just passive, but is an active engagement with the world, a dance that involves both what is out there and what is inside our minds.

This is where the quantum perspective becomes intriguing. Quantum mechanics describes a world where observation itself can influence what is being observed. The classic example is the famous "double-slit experiment," where the behavior of electrons changes depending on whether or not we attempt to observe their path. This suggests that the observer and the observed are not completely separate; they are entangled, interwoven in a very deep way. If the subatomic world operates under this premise, what if our minds are also entangled with the reality around us, such that our consciousness and awareness is an essential element in how we create and interact with the world?

This leads me to wonder, could it be that the very act of sensing, of perceiving the world, is also influenced by

The Creative Code: How Quantum Physics Unlocks Our Brain's Potential

quantum mechanics? Could it be that the way we see colors, hear sounds, or feel textures, are all connected to the strange and sometimes counterintuitive world of quantum physics? If this is true, then the experience of perception isn't a purely passive, deterministic process. Instead, it's a creative interaction between our consciousness and the quantum world around us. And this means that there is a very important subjective element to all of our experiences, that each of us is an active part in creating the universe that we exist in.

Think about how artists, when they are at their most creative, seem to be able to connect with their senses in ways that are very hard for others to understand. Musicians, for example, describe how, in moments of inspiration, they can hear a symphony of possibilities in their mind, almost as if their ears are picking up on subtle notes that are not available to the average person. It's as if their creative mind is somehow accessing a wider range of possibilities than what's available in the purely classical world. Could this increased sensitivity be related to a kind of quantum tuning, a way of attuning one's self to the subtle vibrations and possibilities inherent in the quantum world?

Likewise, visual artists seem to be able to see the world in a completely new light, transforming everyday scenes into works of art that move us deeply. This ability to see beyond the surface, to perceive the interconnectedness of everything around us, has always been considered an ability that is specific to very creative individuals. But what if this ability was available to us all, if only we could learn to tap into that part of ourselves that is sensitive to this deeper interconnectedness? What if our minds, in moments of inspiration, tap into this quantum realm?

This is not to suggest that our experiences are all mere illusions, but rather that the world that we experience is a product of a creative and ongoing interplay between what exists in the outside world and what's happening in our minds. And if quantum mechanics has a key role in this interaction, then perhaps what we are experiencing is a quantum symphony of the senses, a rich and dynamic interplay between the objective reality and our subjective perception. As the physicist Niels Bohr put it,

The Creative Code: How Quantum Physics Unlocks Our Brain's Potential

""Everything we call real is made of things that cannot be regarded as real." This statement points to the fact that our daily experiences are based on deeper principles that are beyond our everyday understanding. That beneath the surface of our daily lives lies a deeper realm of reality that is as magical as it is mysterious.

If this is true, then the journey to understand our creative potential is not just about studying the brain as a classical machine, but about exploring how we, as conscious observers, are intertwined with the quantum nature of reality. We are not passive recipients of data, but active participants in the dance of creation, and every moment that we are awake, we are a part of this continuous and ever-evolving creation.

And if the brain is indeed interacting with the world through the lens of quantum mechanics, this has tremendous implications not just for the way that we see art, but for how we interact with each other, and the way that we understand what it means to be human. It means that the creative spark is not just a localized phenomena, a series of electrical signals within the confines of our skulls, but a much wider, more complex interaction that is connected to everything else around us.

It means that, by exploring the quantum nature of our minds, we may not only discover new ways to enhance our creative abilities but also gain a deeper understanding of how our minds actively participate in creating our experience of reality. Perhaps, our journey to understand our creative potential is nothing more than a journey to discover the nature of our connection to the universe.

And perhaps our minds, just like a quantum computer, are able to explore the multitude of possibilities of experience in the quantum realm, before collapsing into a single, novel, creative solution. This, I find, to be a truly amazing perspective that changes not only how we view creativity, but how we view ourselves and the very universe that we live in.

The Creative Code: How Quantum Physics Unlocks Our Brain's Potential

The Quantum Tapestry of Memory and Imagination

If our senses are the instruments through which we experience the world, memory and imagination are the loom on which we weave our individual realities. We gather data through our eyes, ears, touch, taste, and smell, as we explored in the previous chapter. But it's through our memories, and our ability to conjure up images, sounds, and emotions that aren't present, that we bring these experiences together to create the rich and complex tapestry of our lives. And it's within this tapestry that our capacity for creative thought truly blossoms. This leads me to a very interesting question: could it be that the quantum mechanics, which might be a key element in our perception, are also essential in our capacity for memory and imagination?

Traditional neuroscience views memory as a process of storing and retrieving information from the brain's neural networks. Certain pathways become strengthened over time, enabling us to recall experiences, facts, and skills. Our memories are often seen as recordings, which we then play back whenever we need them. But anyone who has ever struggled to remember something, or has ever experienced the way that a memory changes over time, will know that our memories are not as reliable as a recording. They are not fixed, but are rather malleable and subject to change. And it's in this unreliability that I think the true nature of memory reveals itself.

Likewise, our imagination allows us to create images, ideas, and scenarios that go beyond our direct experience of the world. It's the ability to travel to distant lands in our minds, to design technologies that have yet to exist, and to create art that touches the very core of our humanity. And it's through this ability to imagine that we discover new and unexpected possibilities for ourselves and the world around us. As the celebrated author Terry Pratchett aptly wrote: ""Why do you go away? So that you can come back. So that you can see

The Creative Code: How Quantum Physics Unlocks Our Brain's Potential

the place you came from with new eyes and extra colors. And the people there see you differently, too. Coming back to where you started is not the same as never leaving." This quote highlights the transformative power of imagination, and how seeing the familiar from a new perspective enriches our understanding. It's our ability to travel in our minds, that brings a new perspective to our own lives.

But if the brain is indeed a quantum system, as we've proposed, then memory and imagination become something far more than a simple process of recalling facts or generating images. It could be that memory operates not just as a storage system, but as a dynamic, quantum field where past experiences and present moment all come together, and are continuously being revised, reshaped, and reinterpreted. And what if imagination is, in fact, the ability of our minds to explore a vast and infinite landscape of possibilities within this quantum realm?

Consider, for a moment, the experience of recalling a memory. It's not like playing a video from the past; our memories are often reconstructed, pieced together from fragments of information, shaped by our present emotions and beliefs. And perhaps this reconstruction is not just a function of the classical processes within the brain. Maybe, in the act of remembering, our minds reach into a quantum field of possibilities, exploring different versions of the past before settling on the one that feels most real to us. It's as if we are reaching into the quantum realm and choosing one of the possible realities that we had experienced.

Similarly, when we use our imagination, when we create new art, or develop a new idea, could it be that our minds aren't just generating new thoughts, but are exploring new possibilities within the quantum realm? That is, could it be that our ability to imagine is simply a reflection of the brain's ability to interact with the quantum world, and that this interaction allows us to create completely new ideas and experiences? This is very similar to what a quantum computer does, and therefore, this also leads me to believe that the way our brains operate is perhaps similar to the way quantum computers do.

The Creative Code: How Quantum Physics Unlocks Our Brain's Potential

The act of creativity, from this point of view, becomes a kind of quantum dance, a continuous exploration of both our past and our potential future. We use our memory to draw upon past experiences and we use our imagination to generate new ones. And when this happens, perhaps our minds are interacting with the quantum world, and the possibilities of what can be, become infinite. This interplay between our memories and our imagination, when viewed from a quantum perspective, takes on a whole new and more mysterious dimension.

Perhaps the reason why certain memories seem so vivid, so emotionally resonant, is because they are associated with particular quantum states, and each time that we recall them, our minds are making use of the same quantum pathway. Perhaps there are some specific combinations of neurons that create quantum entanglement, which makes certain memories so easy to remember and other ones so difficult. Perhaps this is why certain memories come back to us with all the details, and yet, other memories remain faint and vague.

Likewise, perhaps the reason why certain people seem to have a more active imagination, a greater creative potential, is that their minds are more adept at accessing and exploring this quantum realm of possibilities. Perhaps through training, focus, and the right type of approach, all of us can learn to tap into this potential, and unlock our own unique creative abilities. And if that is the case, then our capacity for imagination isn't simply something we are born with, but rather, is a potential that exists within us all. As the author Gabriel Garcia Marquez wisely observed: "Memory is not a passive recording of the past but an active creation of the present." This highlights that our memories are always connected to our current experience, and not a static recording. This means that our past, present, and future, are all interconnected.

The idea that memory and imagination are influenced by quantum mechanics might sound far-fetched. However, it also leads to some profound possibilities, not only for how we understand ourselves, but for how we can unlock our full human potential. The journey towards truly understanding our creative code requires us to

The Creative Code: How Quantum Physics Unlocks Our Brain's Potential

explore not only what our brains can do, but also how our minds interact with the quantum world.

And so, as we delve further into this exploration, it becomes clear that the human mind isn't just a container of knowledge, or a processor of information, but also a vibrant, ever-changing quantum landscape where memory and imagination interact to create our very own personal reality. And if we have a deeper understanding of this quantum process, then we can not only learn to understand our past better, but we can also learn to imagine the future that we would like to create.

Kendir Ramiz

The Quantum Connection and Consciousness

Our exploration thus far has taken us deep into the potential quantum nature of computation, perception, memory, and imagination. But the human experience is not just about individual consciousness; it is deeply rooted in our ability to connect with one another, to share our thoughts, emotions, and experiences. This sense of connection is what makes us a species, and perhaps, it is also deeply intertwined with the quantum nature of our minds. This leads me to wonder, are our connections with each other merely the results of brain activities, or is there more to it than that?

The Creative Code: How Quantum Physics Unlocks Our Brain's Potential

Traditional views of human interaction often focus on the exchange of information through language, gesture, and facial expressions. We see each other as individuals, with our own separate minds and experiences. And this perception is so ingrained in us that it feels as if we can never truly understand what's going on in the mind of another person. However, when we look at the world of quantum mechanics, we find something rather different, a completely unexpected interconnectedness. Perhaps, when we try to understand our own human connection from a quantum perspective, we might gain new insights into how our minds are connected with one another.

Think about those moments when you feel deeply connected to another person, a moment where you feel you truly understand what they are going through, almost as if you can read their thoughts. There is a sense of ease, a sense of flow, a sense of harmony that is very difficult to explain using our usual classical way of describing our relationships. Is it possible that in those moments, our minds are experiencing a type of quantum entanglement, where our thoughts, emotions, and perhaps even our consciousness, become intertwined?

The idea of entanglement, as we explored earlier, suggests that two quantum particles can be linked in such a way that they share the same fate, even when separated by vast distances. What if our brains, too, are capable of creating such links, especially during moments of deep connection? This is of course, a very far-reaching idea. But when we see how profoundly connected we are with each other, this doesn't seem as far-fetched. Perhaps our consciousness isn't a purely individual phenomena, but rather, something that is deeply interconnected, and interwoven with other minds.

This possibility also challenges our understanding of consciousness itself. For centuries, consciousness has been seen as something that arises from the activity of the individual brain. But if quantum mechanics has taught us anything, it is that the world is not as separated and distinct as we had once thought. Perhaps consciousness itself, is not a singular event, but a wider phenomenon that is deeply connected to all other

The Creative Code: How Quantum Physics Unlocks Our Brain's Potential

minds. As the philosopher Alan Watts eloquently put it, ""You are a function of what the whole universe is doing in the same way that a wave is a function of what the whole ocean is doing." This profound statement points towards a deep interconnectedness of all things. And perhaps our individual minds are part of a greater cosmic consciousness.

This brings me to the idea that human consciousness might not just be the product of individual brains, but a kind of quantum field that connects us all. Perhaps our individual minds are not as separate as we tend to believe. Perhaps in moments of great creativity, when artists, scientists, or inventors come up with novel ideas, they are tapping into this field of quantum consciousness and drawing from a source of knowledge that is far greater than themselves. And if this is true, then the potential of the human mind is far greater than we tend to imagine.

Consider the power of collective creativity, the way that groups of people can collaborate to create something truly extraordinary. The experience of jamming with other musicians, improvising on stage, can feel as if you are tapping into a shared source of inspiration, something that is far greater than the sum of each individual parts. Could this also be related to a kind of quantum entanglement, a connection that transcends our individual limitations, and creates something truly new and unique?

The power of shared experience and connection is something that we have all felt. The experience of reading a book that moves you deeply, or watching a movie that changes your perspective on life, or listening to a piece of music that seems to speak directly to your soul, are all instances of human connection that go far beyond the simple exchange of information. Perhaps, our brains are tuning to similar quantum states when we participate in shared experiences, and perhaps it is through these shared experiences that we are able to unlock a much deeper understanding of ourselves, others, and the world around us.

And if our human connections are indeed intertwined with the quantum nature of our minds, this has profound

The Creative Code: How Quantum Physics Unlocks Our Brain's Potential

implications for not only how we understand creativity, but also for how we create social structures, governments, and even societies. The potential for humanity becomes infinitely greater when we view each other, not as individuals, but as an extension of our own selves. We are all connected. We all share the same potential, and we can only achieve our highest potential together.

It is in these moments of deep human connection, that I am most convinced that our brains are deeply connected to the quantum world, and that we are not just observers of this world, but active participants in shaping it. And maybe this potential for deep human connection is also related to our potential for creativity, and for the discovery of completely new and innovative ways of thinking and being.

This also means that, the journey to understand our creative potential is not only a journey of our minds, but also a journey of our collective selves. It is a journey that will inevitably bring us closer to each other, and closer to the core of our own humanity. And if we are lucky, we

Kendir Ramiz

will be able to create a more just, a more creative, a more sustainable, and a more beautiful world, where we can all thrive.

The Creative Code: How Quantum Physics Unlocks Our Brain's Potential

Uncertainty and the Creative Process

Our exploration of the quantum brain has taken us to the very edges of what we know about the nature of reality. We've considered the possibility that our thoughts, perceptions, memories, and connections with one another may be influenced by the strange and sometimes paradoxical laws of quantum mechanics. But the more deeply I delve into these possibilities, the more I am confronted by the crucial role that uncertainty plays in the entire creative process. In the quantum world, nothing is fixed, nothing is certain, and it is in this realm of infinite possibilities that we must learn to live, and to create.

Kendir Ramiz

Traditional views of creativity often emphasize planning, structure, and control. We think that in order to create, we must first have a plan, a method, and a well-defined goal. But what if the most innovative and original ideas come from a place of uncertainty, from a willingness to let go of these pre-conceived ideas and venture into the unknown? It's as if, when we attempt to plan every aspect of the creative process, we are placing artificial limits on our own potential, and we might be closing ourselves to some very important possibilities. This reminds me of the words of the brilliant choreographer Martha Graham, who once said "The main thing is to be moved, to love, to hope, to tremble, to live." This quote points to the idea that emotions, and letting go, might be even more important than reason and planning when it comes to creativity. Perhaps, the greatest creativity comes from the heart, and not the head.

In the quantum world, the principle of uncertainty, as we discussed previously, is a fundamental part of reality. We cannot know the position and momentum of a subatomic particle with complete accuracy, and this inherent uncertainty seems to be part of the nature of

The Creative Code: How Quantum Physics Unlocks Our Brain's Potential

the universe. Maybe this concept of uncertainty is also a guiding principle in our creative process, and perhaps the most innovative solutions only emerge when we let go of the need for certainty, when we learn to navigate this quantum labyrinth of endless possibilities.

When we attempt to control every aspect of the creative process, we are limiting our minds to the realm of the known, to what has already been thought, and to what we already expect. Perhaps, if we are to create truly original ideas, then we must allow our minds to operate in a realm of uncertainty, a place where our consciousness becomes fluid and open to new possibilities. It's almost as if the need for control, is precisely what's stopping us from achieving our true creative potential.

Consider those moments when a writer experiences "writer's block," a period where their creative well seems to run dry. This may be because of the fear of the unknown, the fear of not having the right words or ideas. Perhaps, what is needed in these moments, isn't more structure, more planning, but rather, a willingness to embrace uncertainty, to trust that new ideas will emerge, to just sit with the unknown until a creative spark appears from within. It's as if we have to learn to be comfortable in this quantum labyrinth of uncertainty, so that we can experience the joy of creation.

Likewise, in the world of science and technology, the most important breakthroughs often happen when scientists are willing to question existing paradigms, when they are willing to challenge established wisdom, and when they allow their minds to explore a wide range of possibilities without pre-conceived ideas. This requires not only great knowledge and intelligence, but also a capacity to be open to new ideas that come from unexpected places. Perhaps it is precisely in this space of uncertainty, this openness to the unknown, that the truly groundbreaking ideas emerge.

This brings me to consider the importance of play in the creative process. When we play, we are free from the constraints of rigid thinking, we allow our minds to wander, to explore, and to experiment, without worrying about whether we're going to succeed or not. And it may

The Creative Code: How Quantum Physics Unlocks Our Brain's Potential

be that play itself, is a vital mechanism for tapping into the quantum realm of infinite possibility. Perhaps it is through play that we can learn to use uncertainty, and channel it into our creative potential.

The quantum perspective suggests that the creative process isn't a linear, predictable path, but a kind of dance, a constant interplay between certainty and uncertainty. We make use of planning, structure, and methodology, but also allow ourselves to be open to the unexpected, to embrace the unknown, to let go of our need to control every aspect of the creative process. And it's within this dance of the known and the unknown, that we learn to truly unlock our full creative potential.

Perhaps, the key to navigating this quantum labyrinth, is to cultivate a kind of "beginner's mind," an approach that allows us to be open to new possibilities, to approach the world with a sense of wonder and curiosity, and to let go of the need to know everything. For it's in these moments of uncertainty that the greatest discoveries are often made. And if we can learn to be more comfortable with the unknown, then perhaps we can also learn to tap into our full creative potential.

Therefore, the journey to understand our creative code is not just about understanding how the brain functions. It's also about learning to navigate the quantum labyrinth of uncertainty, to embrace the unknown, and to allow our minds to explore the infinite possibilities that exist within ourselves and the world around us. It is within this dance of the known and the unknown, where I believe, our creativity can truly blossom. And if we are open to the possibilities, then the results can be truly remarkable.

The Creative Code: How Quantum Physics Unlocks Our Brain's Potential

The Quantum Pathways to Transformation

Our journey through the potential quantum nature of the mind has led us to explore the realms of perception, memory, imagination, connection, and creativity. But the true value of this exploration lies not just in understanding how our minds work but in discovering how this knowledge can bring about real transformation in our lives. The idea that our brains may be deeply connected to the quantum realm opens up new possibilities for personal growth, for healing, and for unlocking our full human potential. This leads me to consider, how can our understanding of quantum mechanics change the way we lead our lives?

Kendir Ramiz

Traditional approaches to personal transformation often focus on changing our behaviors, our thoughts, and our beliefs. We are encouraged to set goals, to develop new habits, and to challenge our limiting beliefs. While these approaches can be effective, they often feel like we are working on the surface, only dealing with the effects of our problems, and not their underlying causes. Perhaps a quantum perspective can help us access deeper levels of transformation, allowing us to create real, lasting change in ourselves and the world around us.

When we think about the quantum world, we realize that nothing is set in stone, nothing is fixed. Particles can exist in multiple states simultaneously, and our actions as observers play a key role in shaping reality. If we translate this to our own experience, then perhaps this means that our own lives aren't predetermined, that our habits are not fixed, and that we have the ability to create a new and better reality for ourselves. This idea, that we are not fixed, but in a constant state of change, is the starting point of our journey to personal transformation.

The Creative Code: How Quantum Physics Unlocks Our Brain's Potential

Think about those times in your life when you've experienced a true transformation, a moment when you changed the very way you see the world. What allowed you to make that shift? Often it's not just a simple decision, but a profound experience that touches the very core of who we are. Perhaps it is through these intense experiences that we create new neural pathways in our brain, new ways of experiencing the world, that reflect the quantum nature of our minds. And perhaps, what we call 'transformation', is actually a shift in our brain's quantum state.

Perhaps our greatest challenge is that we have been taught to see ourselves as being very limited, as being fixed in our capabilities and our potential. Perhaps, the greatest challenge of our lives, is to see that these limitations are self-imposed, and that we can move beyond them by changing the way that we relate to our world, and to ourselves. As the poet Mary Oliver wrote, ""Tell me, what is it you plan to do with your one wild and precious life?" This question invites us to reflect on how we want to shape our own existence, how we want to use our limited time on this planet to make a difference. Perhaps, our greatest potential lies in our ability to change and grow, and to tap into the infinite possibilities of what could be.

The quantum world also tells us that our actions have ripple effects, that everything is interconnected. If we act from a place of love, compassion, and kindness, we have a positive influence on ourselves and the world around us. Likewise, if we act from a place of fear, anger, and resentment, we are only creating more suffering for ourselves and those around us. And perhaps, if we understand these effects in the quantum realm, we can better chose how we act in the world. And if it's true that we are all interconnected, then perhaps transformation is not an individual journey, but a journey that we must all take together.

This understanding also changes how we view healing, particularly in the context of our physical and mental wellbeing. If our minds are interacting with the quantum realm, then perhaps the process of healing is not just about fixing what is broken, but also about aligning our

The Creative Code: How Quantum Physics Unlocks Our Brain's Potential

quantum state to allow for healing to occur from within. If our physical and mental state is entangled with our quantum state, then perhaps changing the quantum state can have a positive influence on our health and wellbeing.

The quantum pathway to transformation, therefore, involves not only changing our thoughts and beliefs, but also changing the way we interact with the quantum world around us. It invites us to let go of the need for control, to embrace uncertainty, to live from a place of love and compassion, and to align our actions with our highest intentions. It invites us to see ourselves, not as limited, separate individuals, but as interconnected parts of a much larger quantum field. And if we can embrace this quantum perspective, then we can unlock a level of transformation that goes beyond anything we have ever experienced. It is a journey that calls us to be our best selves, and to see our lives as a work of art in progress.

And perhaps the most powerful lesson that we can learn from the quantum world, is that our greatest power lies not in our ability to control things, but in our ability to influence them, to shape them, and to create a new reality for ourselves and the world around us. This potential is not just for a few chosen individuals, but it's available to us all. All it requires is that we open ourselves to the quantum possibilities, and allow ourselves to transform into who we are meant to be. It's a journey that requires courage, curiosity, and, most of all, the deep belief that a better version of ourselves, and our world, is possible.

The Creative Code: How Quantum Physics Unlocks Our Brain's Potential

Shaping a Creative Future

As we approach the conclusion of our exploration, it becomes clear that the journey to understand the quantum nature of mind is not just a scientific endeavor. It's a cultural, philosophical, and deeply personal journey that has profound implications for how we create our future. Understanding the interconnectedness of our minds, our capacity for creative transformation, and our ability to influence reality itself through our actions, puts us in a unique position to shape a future that is more creative, more compassionate, and more aligned with our highest human potential. This leads me to consider, what kind of a future will we be creating now that we have a deeper understanding of ourselves and the world?

Traditional approaches to creating the future have often focused on technology, efficiency, and economic growth. We often imagine a future that is purely based on new technologies that have solved all of our current problems. However, I am starting to believe that a truly meaningful future is not just about technological innovation, but it is about our ability to tap into our creative potential, to build more compassionate communities, and to embrace the interconnectedness of life. And in this process, the quantum view of the world becomes an essential component of how we approach our lives.

If our minds are indeed capable of accessing a quantum realm of possibilities, then this opens up an infinite potential for innovation and creation. Perhaps the solutions to many of our current problems already exist, and it is only through our creative and imaginative abilities that we can discover them. Perhaps our ability to imagine is far greater than we tend to believe, and if only we learn how to fully tap into that potential, then we can find the solutions that we are searching for.

The Creative Code: How Quantum Physics Unlocks Our Brain's Potential

This means that, our educational systems, our workplaces, and our communities should be fostering creativity, encouraging exploration, and nurturing the unique potential within each individual. It isn't simply a matter of teaching our kids a series of facts and information. But rather, we need to be teaching our children how to question, how to imagine, and how to approach the world from a place of curiosity and wonder. And in doing so, we are teaching them how to tap into their own unique potential, and how to become the best versions of themselves.

This also has implications for how we design our social structures and our governments. If we understand that we are all interconnected, then it becomes clear that our social structures must be based on principles of fairness, equality, and justice. We need to create societies that empower all of its members, not just a select few. The quantum view of the world reminds us that we are all part of one human family, and that we can only achieve true greatness if we work together to build a better future for all.

This also means that we need to be more mindful of the impact of our actions on the world around us. We need to move away from the idea that we are separate from nature, and embrace the fact that we are part of the web of life. We need to protect our environment, respect all living beings, and build a more sustainable and harmonious relationship with the planet. As the environmentalist Rachel Carson observed: "The more clearly we can focus our attention on the wonders and realities of the universe about us, the less taste we shall have for destruction." This observation reminds us that it's our connection to nature that can truly inspire us. The more that we learn to appreciate all that is around us, the more we want to protect it for future generations.

The quantum view, also challenges our ideas about what counts as success. Instead of defining success by our material possessions, our social standing, or our economic power, we need to start defining success by our level of creativity, our capacity for compassion, and our contribution to the wellbeing of the planet. If we want to be truly successful, then we must measure success

The Creative Code: How Quantum Physics Unlocks Our Brain's Potential

by the quality of our lives, not the quantity of our possessions.

This shift in perspective also means that we need to move away from fear and competition, and move towards collaboration and love. We need to start seeing each other as partners in the journey of life, and embrace the idea that when one of us succeeds, we all succeed. This vision requires that we create communities where everyone feels valued, where everyone is supported, and where everyone can contribute to the wellbeing of the whole. And this, is the quantum future that I dream of.

The journey towards shaping a creative future is a journey of collective transformation. It requires us to not just understand the quantum nature of the mind, but to integrate this understanding into every aspect of our lives. It's a journey that calls us to be more creative, more compassionate, more connected, and more aware of our responsibility to create a more sustainable future for all. And if we are able to embrace this quantum perspective, then our potential for growth is limitless.

Kendir Ramiz

Embracing the Unknown

Our exploration into the potential quantum nature of the mind has brought us to the very horizon of what we know, and also into the mysterious realm of what is still unknown. We have discovered the possibility that our thoughts, feelings, and perceptions might be influenced by the strange laws of quantum mechanics. We have seen how this knowledge can open up new ways of thinking about creativity, connection, and transformation. But it is also important to acknowledge that there are still many questions that remain unanswered, many mysteries that have yet to be unveiled. This leads me to reflect, what can we learn from the unknown, and how can we embrace the mysteries that still remain?

The Creative Code: How Quantum Physics Unlocks Our Brain's Potential

Traditional approaches to knowledge often focus on what we know, on what has been proven, on what has been measured. The goal is often to find answers, to find certainty, and to make sense of the world. But the quantum world challenges this idea, it reminds us that the universe is infinitely more complex, infinitely more mysterious, and infinitely more surprising than we tend to believe. The more we learn, the more we realize that the more there is that we don't know. Perhaps our greatest journey is not just what we learn, but what we discover that we are yet to learn.

The quantum world also tells us that uncertainty is not an obstacle to knowledge, but rather a vital ingredient to our understanding of reality. Without uncertainty, without the exploration of the unknown, without challenging the status quo, we would be unable to grow, to learn, or to create. And it's this embrace of uncertainty that also allows us to create new possibilities, and new ways of being.

Consider the experience of a scientist on the edge of a discovery. They often start their research with questions, ideas, and maybe even a hunch, that there is something new that is yet to be discovered. They immerse themselves in the unknown, they struggle with their own doubts and uncertainties, and it's only after exploring all these possibilities, that a new breakthrough is finally made. This process of exploration, of pushing the boundaries of our understanding, is at the core of all scientific progress. And perhaps, this ability to embrace uncertainty, is also the key to unleashing our own creative power.

Likewise, when we look at the history of art and creativity, we realize that the most influential and lasting pieces of art are often the ones that challenge our perceptions, that invite us to see the world in a completely new way, and that push the boundaries of what's possible. Artists, when they are at their best, aren't afraid to venture into the unknown. They are willing to explore the infinite possibilities that exist in their imaginations, and they use their work to help us see things that we have previously not seen.

The Creative Code: How Quantum Physics Unlocks Our Brain's Potential

Perhaps, the greatest lesson that we can learn from the quantum world, is that it's ok not to have all the answers. It's ok to live in the space of uncertainty, and that it is in fact, in this space of not knowing, that the greatest growth, transformation, and creative potential can emerge. The human mind can only grow, if we allow it to venture into the unknown. And we can only truly understand ourselves, if we are willing to embrace our own uncertainty.

This also means that, the journey to understand our creative potential, is not something that is going to be completed. It's a continuous journey, a process of exploration that has no end. The more we learn about the quantum world, the more we learn about ourselves, and the more we realize, that there is always more to learn. As the Nobel Prize winner Rabindranath Tagore once wrote "The butterfly counts not months but moments, and has time enough." This observation reminds us that time is not a fixed entity, but something that is shaped by our experience. And this experience, is always shifting and evolving.

Therefore, our quest to understand the quantum nature of our minds is not about finding the definitive answers. But rather, it's about learning to live with the questions, to embrace the mystery, and to explore the infinite possibilities that lie within ourselves and the world around us. It's an invitation to see life, not as something that is fixed, determined, and predictable, but as a continuous and ever-evolving dance of creativity, and transformation.

As we stand at the quantum horizon, we should not only marvel at the discoveries we've made but embrace the infinite possibilities that still remain unexplored. The true beauty of the quantum world lies not just in what we know, but in what we have yet to discover. And that, I believe, is the essence of our human journey. A journey of infinite possibilities, where we are constantly shaping, not only ourselves, but the world that we live in. This journey never truly ends, and that is where it's true beauty resides.

The Creative Code: How Quantum Physics Unlocks Our Brain's Potential

The Quantum Symphony of Becoming

Our journey through the potential quantum nature of the mind has reached a point where the lines between science, philosophy, personal experience, and creative vision all begin to blur. We've explored the possibility that our thoughts, perceptions, memories, connections, and even our deepest sense of self are intertwined with the strange and sometimes paradoxical laws of quantum mechanics. And as we come to the end of this exploration, it becomes clear that this is not just a journey to understand our brains, but a journey to understand our very place in the universe. This brings me to a place of profound contemplation: how can we integrate the lessons of quantum mechanics into our everyday lives, and how can we use this knowledge to navigate the complex and ever-changing world around us?

Traditional views of the universe often portray it as a mechanical, deterministic system, governed by fixed laws that operate in a predictable manner. The very idea that the universe operates like a clock, that if you know the position and momentum of all the objects, that you can predict the future, has been at the very core of our understanding for hundreds of years. But quantum mechanics challenges this view, it shows us that the universe is far more fluid, more dynamic, and more open to possibilities than we ever imagined. This perspective, that the universe is not fixed, but rather, a continuous process of becoming, has tremendous implications for how we see ourselves, and our potential for the future.

If our minds are indeed influenced by the quantum world, then it also means that we are not just passive observers of our reality, but active participants in shaping it. We are not just the products of our environment, but also the creators of our own experiences. This means that our thoughts, feelings, beliefs, and our actions, are continuously influencing the reality that we experience. And if we are to take responsibility for our lives, then we must also embrace this responsibility of shaping our lives based on how we view our world. The world that we see around us is a mirror of what is going on within our minds, and this connection to our world, requires that we become more mindful of what we are creating.

This brings me to consider how the idea of a quantum brain challenges the very notion of 'self' itself. We often perceive ourselves as separate, autonomous beings, with our own unique thoughts, emotions, and desires. But the quantum perspective suggests that we are far more interconnected, that our individuality is intertwined with the greater whole. Perhaps our very sense of 'self', is an illusion, and that we are not individuals, but rather, expressions of one universal consciousness that pervades all of reality. As the philosopher Carl Jung observed: "The meeting of two personalities is like the contact of two chemical substances: if there is any reaction, both are transformed." This observation reminds us that we are all influenced by our interactions with each other, and therefore, the notion of a singular self may not be a completely accurate view of reality.

This possibility also challenges our usual way of thinking about what it means to be human. If we are not just individuals, but rather interconnected parts of a quantum field of consciousness, then this puts a new responsibility on us to act in ways that are not only beneficial to ourselves, but also beneficial to others. What this view calls for is a profound shift in our values, our ethics, and the way that we engage with the world. It calls for a transition from a purely self-centered viewpoint, to one where we are concerned with the well-being of all. It calls for a transformation of our own hearts and minds.

But this change, is something that I believe to be not only essential, but also, within our reach. If we are to take responsibility for our lives, then we must embrace our connection to the whole, and act in ways that are guided by love, compassion, and a deep respect for all life. If our minds are entangled with everything around us, then we must choose to engage in ways that are constructive, creative, and aligned with the very best of human potential. This, is the vision that I have for the future of humanity. A future where we live in peace with each other, and in harmony with the planet.

The Creative Code: How Quantum Physics Unlocks Our Brain's Potential

The exploration of the quantum mind is not just a scientific quest; it's an artistic endeavor, a creative project in which each of us is invited to participate. The implications of a quantum influenced mind touch on every aspect of our lives, and it's up to each of us to fully understand these implications and act accordingly. This means that, we must not just be concerned with the intellectual understanding of quantum principles, but we must also integrate these insights into our values, our actions, and our choices. It's a journey that calls for our full engagement, our whole heart, and our deepest compassion.

Traditional knowledge systems often portray the world as fixed and unchanging. But quantum mechanics reveals a world that is always in motion, always in flux, always evolving. The very idea that the universe is in a constant state of becoming, challenges all that we thought we knew about reality, and this idea, also challenges our understanding of ourselves, and our potential for growth. If our reality is ever-evolving, then our minds must also be ever-evolving to meet the challenges of this ever-changing reality.

This brings me to consider that the journey to understand the quantum nature of the mind is not something that will ever be completely finished. It's not about finding all the answers, or solving all the mysteries. It's rather, a continuous exploration, an ongoing dance between the known and the unknown. The more we discover, the more we realize, the more that there is still to be understood. And this journey, has to be ongoing, with each generation of humanity learning from the past, and exploring all the possibilities of the future.

Perhaps the most important lesson that quantum mechanics can teach us, is that the universe is not something that is separate from us. It is not something 'out there', but rather something that we are intimately connected to. We are not just observers of the world, but participants in its ongoing creation. And what we think, feel, and believe has a direct impact on the world that we experience. And perhaps, the true power of the

The Creative Code: How Quantum Physics Unlocks Our Brain's Potential

human mind, lies in our ability to shape reality according to our collective will. As the visionary Buckminster Fuller wisely said, ""You never change things by fighting the existing reality. To change something, build a new model that makes the existing model obsolete." This statement points to the need to create new possibilities, instead of fighting against current ones.

This understanding also challenges us to think about the future in a completely new way. If our thoughts, feelings, and beliefs are constantly shaping reality, then we must use this power wisely. We need to use our creativity to build a future that is aligned with our highest ideals. It is no longer enough for us to just passively exist in the world. We must instead become actively engaged in creating the world that we want to see.

Therefore, the journey to understand our creative potential is not just about personal growth. It's about our collective evolution as a species. It is a journey to build a more beautiful, a more sustainable, and a more compassionate world, where all life can flourish. The future of humanity, depends on us.

As we conclude this exploration, I am filled with a deep sense of wonder, and a strong belief in our capacity for creative transformation. The journey that we have taken is not just a reflection on the possibility that our minds are entangled with the quantum world. It's also a call for action, an invitation to fully embrace our potential, and to take responsibility for the future of our shared world. And it is within this context, that our individual lives can take on a whole new level of purpose.

Traditional endings often portray finality, completion, and resolution. But life is not about finality. It is a continuous and ever-evolving journey that never truly ends. Just as the universe itself is constantly expanding, so too is our knowledge, our understanding, and our potential for growth. What we consider to be the end, is in reality, just another beginning. The end of one chapter, is simply the start of another. And it is by recognizing this constant state of change that we can truly embrace the nature of our lives.

The Creative Code: How Quantum Physics Unlocks Our Brain's Potential

Perhaps the most profound lesson that we can take away from the quantum world is the importance of embracing the unknown. When we allow ourselves to venture into uncharted territories, and when we embrace the uncertainties of life, then we open ourselves to new possibilities, and we create the conditions for truly meaningful transformations to occur. It is often by entering into the darkness, that we truly discover the light.

Therefore, the journey of understanding the quantum mind isn't something that is just for scientists, or philosophers, or artists. It's a journey for all of humanity. It's a journey to understand our own potential, and our ability to shape the world around us. It is a journey that requires courage, curiosity, and a deep belief in the possibility of a better world. As the writer Henry Miller once said "Our destination is never a place, but a new way of seeing things." And perhaps, it is in this new way of seeing, that we can find the true meaning of our own journey, and our role in the ongoing creation of our universe.

Kendir Ramiz

The understanding of the quantum nature of our mind is not a final destination. It's a pathway to a deeper understanding of ourselves, our connection to each other, and our place in the universe. It is an invitation to live our lives with more creativity, more compassion, and more awareness of the infinite possibilities that lie within us.

It's a reminder that our lives, much like a quantum system, are a symphony of interconnected elements, constantly evolving, always open to new possibilities. And each of us, is invited to participate in this ongoing symphony, to play our unique part, and to contribute to the beauty and wonder of the world. Let the music continue.

www.ingramcontent.com/pod-product-compliance
Lightning Source LLC
Chambersburg PA
CBHW070357230526
45471CB00006B/2613